Janice VanCleave's
WILD, WACKY, AND WEIRD
SCIENCE EXPERIMENTS

Many More of Janice VanCleave's
Wild, Wacky, and Weird
PHYSICS EXPERIMENTS

Illustrations by
Jim Carroll

New York

This edition published in 2018 by
The Rosen Publishing Group, Inc.
29 East 21st Street
New York, NY 10010

Introduction and additional end matter copyright © 2018 by The Rosen Publishing Group, Inc.

All rights reserved. No part of this book may be reproduced in any form without permission in writing from the publisher, except by a reviewer.

Library of Congress Cataloging-in-Publication Data

Names: VanCleave, Janice Pratt, author. | VanCleave, Janice Pratt. Janice Vancleave's wild, wacky, and weird science experiments.
Title: Many more of Janice VanCleave's wild, wacky, and weird physics experiments / Janice VanCleave.
Description: New York : Rosen Publishing Group, 2018 | Series: Janice VanCleave's wild, wacky, and weird science experiments | Includes bibliographical references and index. | Audience: Grades 5–8.
Identifiers: LCCN 2017012201| ISBN 9781499439571 (library bound) | ISBN 9781499439557 (pbk.) | ISBN 9781499439564 (6 pack)
Subjects: LCSH: Physics—Experiments—Juvenile literature.
Classification: LCC QC25 .V2554 2018 | DDC 530.078—dc23
LC record available at https://lccn.loc.gov/2017012201

Manufactured in China

Illustrations by Jim Carroll

Experiments first published in *Janice VanCleave's 203 Icy, Freezing, Frosty, Cool, and Wild Experiments* by John Wiley & Sons, Inc. copyright © 1999 Janice VanCleave and *Janice VanCleave's 200 Gooey, Slippery, Slimy, Weird and Fun Experiments* by John Wiley & Sons, Inc. copyright © 1992 Janice VanCleave

Contents

Introduction 4
Straw Flute 8
Clucking Chicken 10
Moving 12
Immovable 14
Scale 16
Straight Up 18
Slanted 20
Lifter 22
Longer 24
Second-Class Lever 26
Ringer 28
Movable 30
Weightless 32
Taller 34
Down? 36
Paper Weight 38
Reversed 40
Lifter 42
Free-Fall 44
Magic Box 46
Balancing Point 48
Unbalanced 50
Balloon Rocket 52
Dancers 54
   Glossary 56
   For More Information 58
   For Further Reading 61
   Index 63

Introduction

Physics is the study of energy, matter, and forces and their relationship with each other. Physicists study everything from tiny atomic particles to the whole universe! Albert Einstein (1879–1955) is perhaps the most famous physicist, and many choose to follow in his footsteps.

People who decide to work in the field of physics have a variety of career choices. Some engineers investigate aircraft accidents and others work with nuclear power. Some physicists design high-speed trains, roller coasters, or lasers for surgery. All of these people have something in common: they are constantly asking questions to learn even more about physics.

This book is a collection of science experiments about physics. How does a simple machine make work easier? What is the effect of gravity on a person's height? How does a rocket move? You will find the answers to these and many other questions by doing the experiments in this book.

How to Use This Book

You will be rewarded with successful experiments if you read each experiment carefully, follow the steps in order, and do not substitute materials. The following sections are included for all the experiments.

- » **PURPOSE:** *The basic goals for the experiment.*

- » **MATERIALS:** *A list of supplies you will need.* You will experience less frustration and more fun if you gather all the necessary materials for the experiments before you begin. You lose your train of thought when you have to stop and search for supplies.

- » **PROCEDURE:** *Step-by-step instructions on how to perform the experiment.* Follow each step very carefully, never skip steps, and do not add your own. Safety is of the utmost importance, and by reading the experiment before starting, then following the instructions exactly, you can feel confident that no unexpected results will occur. Ask an adult to help you when you are working with anything sharp or hot. If adult supervision is required, it will be noted in the experiment.

- » **RESULTS:** *An explanation stating exactly what is expected to happen.* This is an immediate learning tool. If the expected results are achieved, you will know that you did the experiment correctly. If your results are not the same as described in the experiment, carefully read the instructions and start over from the first step.

- » **WHY?** *An explanation of why the results were achieved.*

Introduction

The Scientific Method

Scientists identify a problem or observe an event. Then they seek solutions or explanations through research and experimentation. By doing the experiments in this book, you will learn to follow experimental steps and make observations. You will also learn many scientific principles that have to do with physics.

In the process, the things you see or learn may lead you to new questions. For example, perhaps you have completed the experiment that studies how air affects the falling rate of objects. Now you wonder what effect the size or shape of the object has on the falling rate. That's great! All scientists are curious and ask new questions about what they learn. When you design a new experiment, it is a good idea to follow the scientific method.

1. Ask a question.
2. Do some research about your question. What do you already know?
3. Come up with a hypothesis, or a possible answer to your question.
4. Design an experiment to test your hypothesis. Make sure the experiment is repeatable.
5. Collect the data and make observations.
6. Analyze your results.
7. Reach a conclusion. Did your results support your hypothesis?

Many times, the experiment leads to more questions and a new experiment.

Always remember that when devising your own science experiment, have a knowledgeable adult review it with you before trying it out. Ask him or her to supervise it as well.

Straw Flute

PURPOSE To determine if the length of a flute affects the pitch of the sound it produces.

MATERIALS drinking straw
ruler
scissors

PROCEDURE

1. Make a ½-inch (1.3-centimeter) cut on each side of the straw's end. This forms the reed part of the flute.

2. Place the reed in your mouth.

3. Push on the reed with your lips and blow. You may have to try several times and change the pressure of your lips in order to produce a sound.

4. As you play the straw flute, cut the end of the straw off with the scissors and observe any change in pitch.

RESULTS The pitch of the sound gets higher as the length of the straw decreases.

WHY? The sound produced is due to the vibration of the straw and the air inside it. The longer the column of vibrating air inside the tube, the lower the pitch of the sound.

Straw Flute

Clucking Chicken

PURPOSE To use a vibrating string to produce a sound.

MATERIALS
    pencil
    paper cup, 6.4 ounces (192 milliliters)
    kite string, 24 inches (60 cm)
    toothpick
    rectangular kitchen sponge
    water

PROCEDURE

1. Use the pencil to punch two holes about ½ inch (1.5 cm) apart in the bottom of the cup.

2. Push the string through the holes and tie it on the outside of the cup.

3. Insert the end of the string in one of the holes and pull it through so that the string hangs out of the cup.

4. Place a toothpick under the loop of string on the outside of the cup with the ends of the toothpick extending over the edges of the cup.

5. Cut a 1-inch x ½-inch (2.5 cm x 1.3 cm) section from the sponge.

6. Tie the end of the string around the center of the piece of sponge.

7. Wet the sponge with water.

8. Wrap the wet sponge around the top of the string.

9. Squeeze the sponge against the string as you move the sponge down the string using jerky movements.

RESULTS A sound is produced like that of a clucking chicken.

WHY? The water allows the sponge to move down the string, but there is enough friction to cause the string to vibrate because the sponge skips and pulls at the string. This irregular touching on the string produces tiny taps that force the string's molecules to move back and forth. The vibrating string strikes the molecules in the cup, and the cup's molecules strike the air molecules, causing them to move back and forth in rhythm with the cup and string. The sound is made louder because the inside of the cup acts like a megaphone that concentrates the sound waves and sends them out in one direction.

Moving

PURPOSE To demonstrate inertia of a moving object.

MATERIALS sharpened pencil
walnut-size ball of modeling clay
marker
ruler

Procedure

1. Insert the point of the pencil into the clay ball. Use the marker to draw a line on the pencil where it meets the clay.

2. Holding the pencil vertically with the clay ball on top, raise the pencil about 2 inches (5 cm) above a table and firmly hit the eraser end of the pencil against the table seven to eight times.

3. Draw a second line on the pencil. Then remove the pencil from the clay and observe how much of the pencil was inserted into the clay.

RESULTS More of the pencil is in the clay after hitting the pencil on the table, and the clay covers the mark on the pencil.

WHY? The pencil and clay ball have inertia, a property of matter that causes objects to resist any change in motion. Because of inertia, objects tend to stay still if they are still and continue to move if they are moving. Only an outside force can change the inertia of an object. When the pencil and clay ball are moving, both have inertia of motion. Hitting the pencil against the table applies a force against the pencil, causing it to stop moving. But the inertia of the clay ball keeps it moving forward for a short time. So the clay pushes against the pencil point, causing the pencil to be pushed farther into the clay. The more securely the ball is attached to the pencil, the less the clay will move.

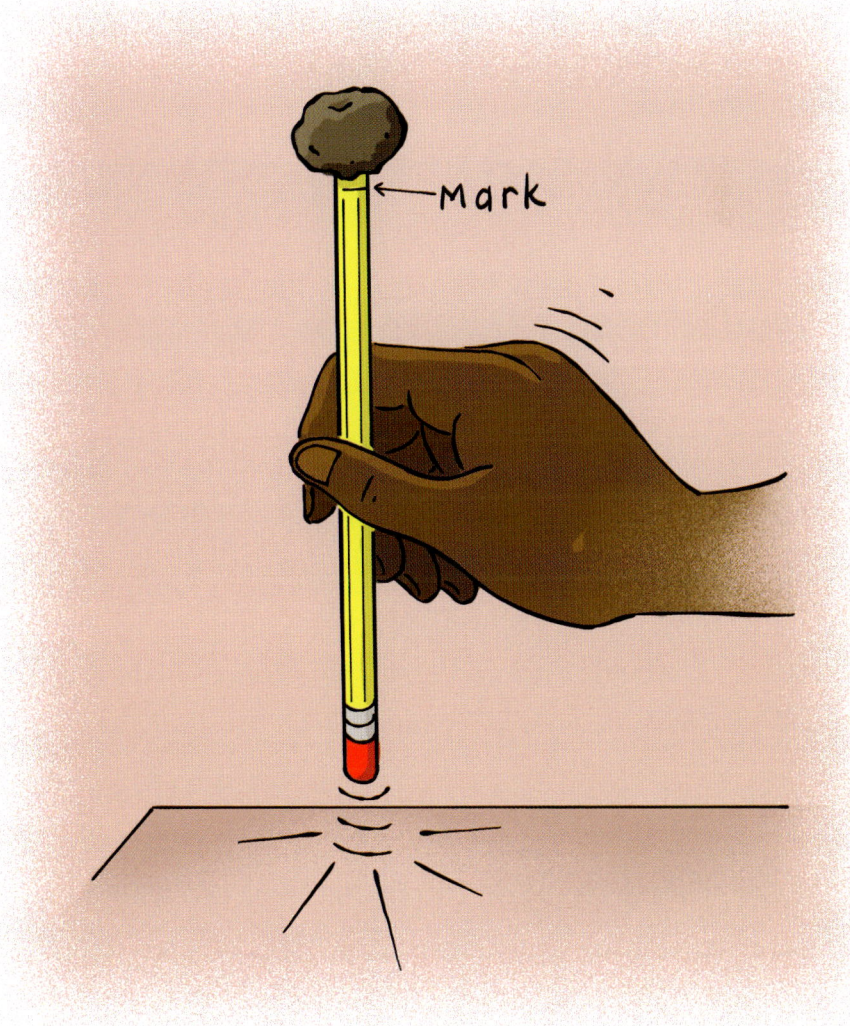

IMMOVABLE

PURPOSE To demonstrate inertia of a still object.

MATERIALS fine-point black marker
2 pencils, 1 sharpened and with a new flat eraser
walnut-size ball of modeling clay
arrowhead eraser

PROCEDURE

1. Use the marker to make a thin line just above the graphite of the sharpened pencil.

2. Insert the point of the marked pencil into the clay ball so that the mark is just above the clay.

3. Place the arrowhead eraser on the other pencil.

4. Hold the marked pencil vertically in midair with the clay ball at the bottom. As you hit the flat eraser of this pencil with the arrowhead eraser of the other pencil fifteen to twenty times, observe the marked area of the pencil.

RESULTS The black mark is not visible, and more of the pencil is inserted in the clay.

WHY? Stationary objects will not move unless a force is applied to them. The pencil is being hit, but the clay ball is not. So the pencil moves into the stationary clay ball.

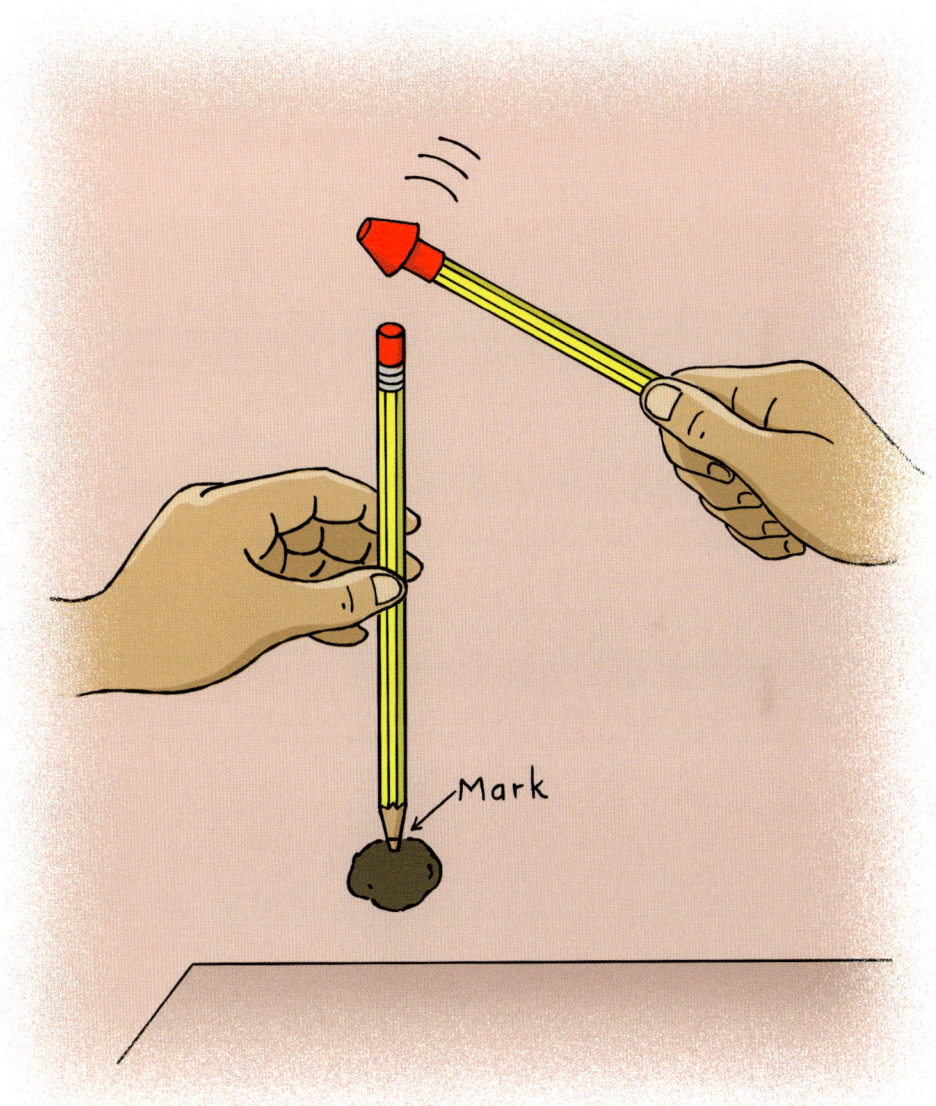

15

Immovable

SCALE

PURPOSE To make a scale.

MATERIALS large rubber band
ruler
masking tape
paper clip
sharpened pencil
5-ounce (180 ml) paper cup
40 to 50 pennies (Any coins or small rocks will work.)

PROCEDURE

1. Secure one end of the rubber band to the back of the zero end of the ruler with tape. Flip the other end of the rubber band over the face of the ruler.

2. Bend the paper clip into a hook. Attach one end of the hook to the rubber band.

3. Use the pencil to make a small hole beneath the rim of the paper cup.

4. Place the hook through the hole in the paper cup.

5. Hold the ruler vertically and observe the ruler measurement at the bottom of the rubber band.

6. Add coins to the cup, ten to fifteen at a time, until all the coins are used. Observe the ruler measurement at the bottom of the rubber band after each addition.

NOTE: Keep the scale for the next experiment.

RESULTS The rubber band stretches, and the ruler measurement increases as more coins are added to the cup.

WHY? Gravity pulls things toward the center of Earth. Thus, gravity pulls the cup and the attached rubber band down. Coins have mass, which causes them to have weight in Earth's gravity. As more coins are added to the cup, the weight of the cup increases and the cup has a greater downward force. Thus, the cup pulls more on the rubber band, which stretches more. While your scale cannot give exact measurements, it can be used to compare the weights of objects.

Straight Up

Purpose To determine the effort force needed to lift an object straight up.

Materials

3 books
1 cup (250 ml) uncooked rice
sock
12-inch (30 cm) piece of string
scale from previous experiment, see Scale pp.16–17
writing paper
pencil

Procedure

1. Stack the books on a table.
2. Pour the rice into the sock, and tie a knot in the sock.
3. Tie one end of the string to the end of the rubber band on the scale and the other end of the string to the top of the sock.
4. Place the sock on the table, and lift the scale straight up until the bottom of the sock is level with the top of the stack.
5. Record the distance the rubber band stretches along the ruler.

NOTE: Keep these materials for the next experiment.

Results The rubber band stretches when used to lift the sock straight up. The distance the rubber band stretches varies depending on the type of rubber band used.

WHY? Work is done when a force is used to move an object. The force you apply to do work is called effort force. In this experiment, gravity pulls the sock down, thus pulling the attached rubber band down. The weight of the sock is the measure of the gravitational force pulling it down. The distance the rubber band is stretched is equal to the distance any equal weight object would cause the rubber band to stretch. The amount of effort force needed to lift the sock is equal to the sock's weight.

Straight Up

Slanted

Purpose To determine the effort force needed to lift an object using an inclined plane.

Materials yardstick (meterstick)
  setup from previous experiment, see pp. 18–19

Procedure

1. Place one end of the measuring stick on the edge of the books to make a ramp.

2. Place the sock on the bottom of the ramp.

3. Holding on to the scale, slowly pull the sock to the top of the ramp by sliding the scale up the ramp.

4. Observe the distance the rubber band stretches and compare it to the distance it stretched in Straight Up.

Results The rubber band stretches a shorter distance when used to pull the sock up the ramp than when lifting it straight up.

Why? A machine is a device that makes work easier by changing the speed, direction, or amount of effort force applied. The ramp is a simple machine called an inclined plane (a flat, slanting surface). It is used to raise an object with less effort force than it takes to raise the object straight up. The fact that the rubber band stretched less indicates that less effort force was needed to raise the sock up the ramp than was needed in the experiment Straight Up to lift it straight up. Whether lifting the sock straight up or pulling it up the ramp, the results are the same: the sock is raised to the height of the books. It just takes less effort force when a machine is used to help do the work.

LIFTER

PURPOSE To demonstrate a first-class lever.

MATERIALS table
sturdy chair (The back should be as tall as the table.)
ruler
broom

PROCEDURE

1. Place the back of the chair about 12 inches (30 cm) from the edge of the table.

2. Lay the broom handle over the back of the chair and under the edge of the tabletop.

3. Place your hand on the straw end of the broom and gently push down.

NOTE: Keep these materials for the next experiment.

RESULTS The straw end of the broom moves down and the other end rises, lifting the table.

WHY? A lever is a machine made of a rigid bar that turns around a pivot point, called a fulcrum, and is used to lift or move an object called the load. A first-class lever is a lever in which the fulcrum is between the two ends of the bar. This type of lever changes the direction of the force applied—one end of the lever moves up when the other is pushed down. It also generally requires less effort force to move a load with a first-class lever than with other types or without a lever.

LONGER

PURPOSE To determine how the position of the fulcrum of a first-class lever affects effort force.

MATERIALS setup from previous experiment Lifter, see pp. 22–23
yardstick (meterstick)

PROCEDURE

1. Repeat the procedure from Lifter, noting how much effort it takes to raise the table.

2. Repeat the procedure two more times, first with the chair 24 inches (60 cm) and then 36 inches (90 cm) from the table. Note the effort it takes each time.

RESULTS The farther away the chair is from the table, the harder it is to lift the table.

WHY? With a first-class lever, less effort force is used when the effort arm (the distance from the fulcrum to the point where you apply the effort force) is longer than the load arm (the distance from the fulcrum to the load). Thus, it was easier to raise the table (the load) when the chair (the fulcrum) was closer to the table.

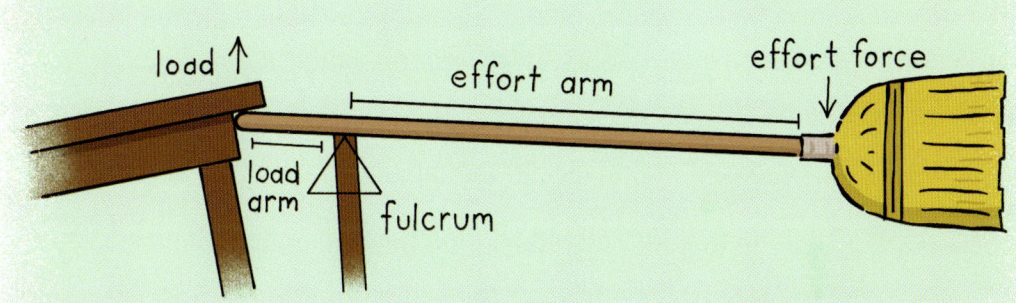

Second-Class Lever

Purpose To demonstrate a second-class lever.

Materials 1-yard (1 m) piece of string
brick
yardstick (meterstick)
masking tape

Procedure

1. Tie one end of the string around the brick, and place the brick on the floor.

2. Holding the free end of the string, lift the brick about 6 inches (15 cm) above the floor. Note the effort required to lift the brick.

3. Place about 2 inches (5 cm) of the measuring stick on the edge of a table. Tape the end of the stick to the table.

4. Set the brick on the floor directly beneath the midpoint of the stick. Lower the free end of the stick, and tie the free end of the string around the center of the stick.

5. Lift the free end of the stick until the brick is about 6 inches (15 cm) above the floor. Again note the effort required to lift the brick.

Results It takes less effort to lift the brick by raising the end of the stick than by lifting it with the string.

Why? The measuring stick acts as a kind of simple machine called a second-class lever. A lever is considered second class when load (the object being moved—in this case, the brick) is between the fulcrum and the effort force (the force you applied to the stick). A second-class lever does not change the direction of the force; the load moves in the same direction as the effort force (the brick moved in the same direction as your hand—up). This type of lever, while not as effort saving as a first-class lever, still requires less force to raise the load than if you lifted it without the lever.

Second-Class Lever

Ringer

Purpose To demonstrate a third-class lever.

Materials 1-yard (1 m) piece of string
yardstick (meterstick)
metal screw ring for 1-quart (1-liter) canning jars
   (Any metal ring with a 2.5-inch (6.4 cm) diameter
   will work.)
glass soda bottle

Procedure

1. Tie one end of the string to the end of the yardstick (meterstick). Tie the free end of the string to the metal ring.

2. Stand the soda bottle on the floor.

3. Wrap one hand around the bottom of the stick, and place the hand you write with immediately above the first hand in the same way you would hold a baseball bat.

4. Stand so that the metal ring dangles directly above the top of the soda bottle.

5. Try to hook the ring over the mouth of the soda bottle by moving the measuring stick with the hand on top only.

Results It is difficult to move the opposite end of the stick small distances; thus it is hard to hook the ring.

Why? The measuring stick is used as a lever, with your writing hand applying the effort force while your other hand marks the point of the fulcrum. The ring is the load that is raised up and down. This is an example of a third-class lever. With a third-class lever, the effort force is always greater than the load force. Generally, the advantage of using a third-class lever is that it multiplies the distance of the effort force. The effort force needs to move only a small distance to move the load a large distance, but in this experiment, it is a disadvantage to use a third-class lever because the load only needs to be moved a small distance.

Movable

Purpose To model the effect of using a movable pulley to lift a load.

Materials 12-inch (30 cm) piece of string
metal spoon
paper clip
masking tape
36-inch (1 m) piece of string

Procedure

1. Attach one end of the short string to the handle of the spoon, and tie the other end of the string to the paper clip.

2. Tape one end of the long string to the edge of a table. Run the string through the paper clip.

3. Raise the spoon by lifting up on the free end of the long string. Note the effort required and the direction the paper clip moves.

Results The spoon is easily moved and moves in the same direction that you pull the string.

Why? A pulley is a type of simple machine that normally is made up of a grooved wheel that turns by the action of a rope or belt in the groove. The paper clip models the effect of a pulley in lifting a load. Since the paper clip is attached to the load (the spoon) and moves in the direction of the effort force, it models a movable pulley. This type of pulley multiplies the effort force, thus making it easier to move load.

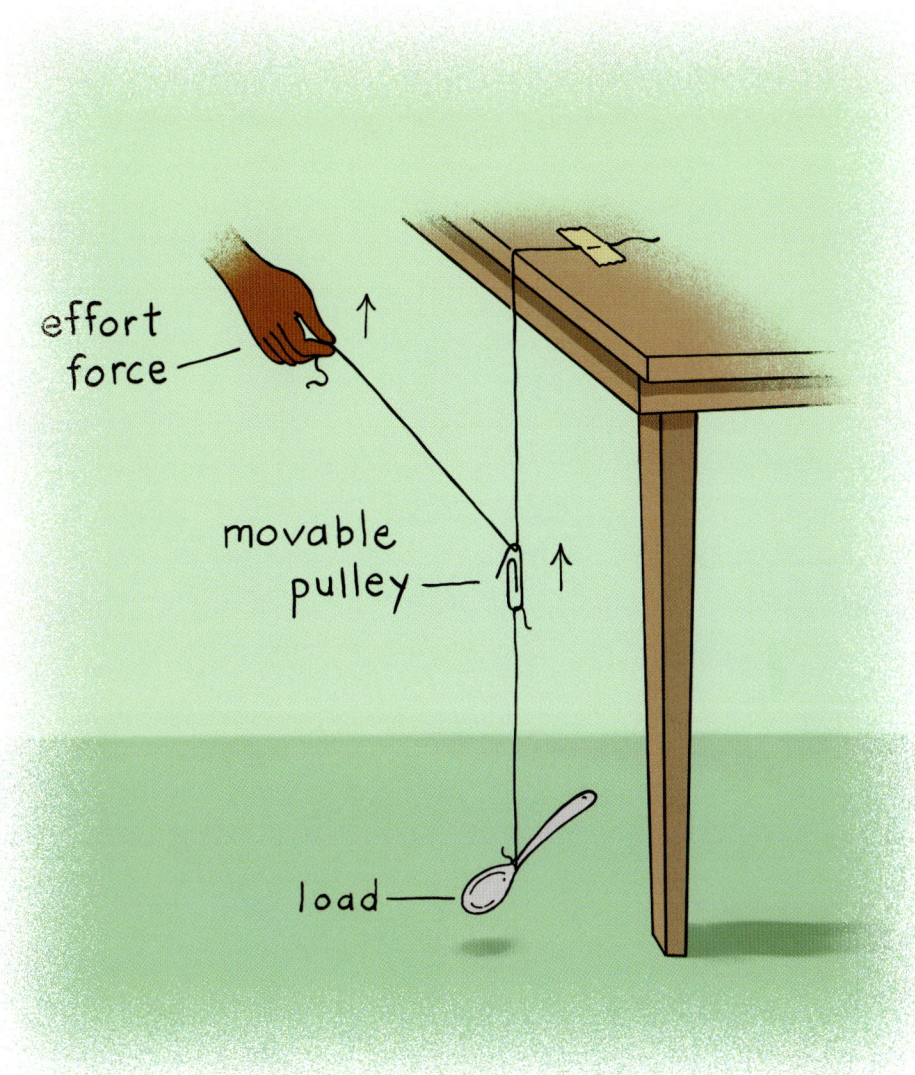

Movable

WEIGHTLESS

PURPOSE To demonstrate apparent weightlessness.

MATERIALS sharpened pencil
9-ounce (270 ml) paper cup
1-quart (1 l) or larger plastic pitcher
tap water
helper

PROCEDURE

NOTE: This activity should be performed outdoors in an area where water can be spilled. You may get wet when performing this experiment.

1. Use the pencil to punch a hole in the side of the paper cup near its bottom.
2. Fill the pitcher with water.
3. Ask your helper to hold the cup so that the hole points away from his or her body. Have your helper hold a finger over the hole as you fill the paper cup with water from the pitcher.
4. While sitting to the side of the cup, observe what happens when your helper removes his or her finger from the hole.
5. Again, have your helper cover the hole. Refill the cup.
6. As you watch from your sitting position, ask your helper to raise the cup as high as possible, then drop it.

RESULTS When the cup is held stationary, the water streams out of the hole. But no water flows out of the hole while the cup is falling.

WHY? Gravity pulls the water down out of the hole in the stationary cup. When the cup is released, gravity pulls both the cup and the water down. The cup and water fall together, thus the water does not flow out of the hole. Objects that fall only due to the pull of gravity are said to free-fall. During a free fall, there is an apparent weightlessness (a state of having no weight). Astronauts in spacecraft orbiting Earth are free-falling because the craft and its contents are constantly falling around Earth. Astronauts and their spacecraft, like the cup and water, fall together and, therefore, cannot push against one another. Thus, they free-fall and the astronauts experience apparent weightlessness.

TALLER

PURPOSE To simulate the effect of gravity on a person's height.

MATERIALS scissors
9-inch (23 cm) round balloon
small baby food jar
1-quart (1 l) widemouthed jar

PROCEDURE

1. Cut the balloon in half.

2. Stretch the bottom of the balloon over the baby food jar to cover its opening.

3. Place the covered baby food jar inside the larger jar.

4. Stretch the top half of the balloon over the mouth of the large jar so that the neck of the balloon is centered over the jar's mouth.

5. Push the stretched balloon down into the jar, allowing air from inside the jar to escape through the neck of the balloon.

6. Twist the balloon's neck, then pull it upward and observe the stretched balloon over the mouth of the baby food jar.

RESULTS The balloon bulges upward.

WHY? The jars are used to simulate the effect of gravity on the movable disks in a person's spinal column. Pulling the balloon upward represents zero gravity, as shown when the rubber covering on the baby food jar bulges upward. In orbit, the height of astronauts increases because they are free-falling, which produces an effect of zero gravity on their bodies. On Earth, gravity pulls a person toward the center of Earth, but Earth pushes back, thus, the disks in the spinal column are pressed together. In free fall, gravity is pulling, as before, but nothing is pushing back. The amount of gravity decreases with the altitude of the spacecraft. Thus, the disks separate, resulting in an increase in height. Skin and other body parts restrict the amount of disk separation.

Taller

Down?

PURPOSE To show that gravity always pulls a free-hanging object down.

MATERIALS paper clip
12-inch (30 cm) piece of string
ruler
2 large books of equal height

PROCEDURE

1. Tie the paper clip to one end of the string. Tie the free end of the string securely around the middle of the ruler.

2. Stand the books about 10 inches (25 cm) apart on a flat surface.

3. Support the ends of the ruler on the tops of the books. Observe the position of the string and paper clip.

RESULTS The paper clip hangs straight down.

WHY? Earth's gravity is the force that pulls objects toward the center of Earth. Thus, gravity pulls the free-hanging paper clip straight down. Down is toward Earth's center.

Down?

Paper Weight

PURPOSE To determine if weight changes the falling rate of objects.

MATERIALS pencil
2 coins, 1 large and 1 small
printer paper
scissors
yardstick (meterstick)

PROCEDURE

1. Use the pencil and the small coin to draw a circle on the paper. Cut out the paper circle.

2. Hold the paper circle under the larger coin so that both are parallel with the floor. Be sure the paper does not extend past the edge of the coin.

3. Raise the coin and paper about 3 feet (1 m) above the floor, then release and allow both coin and paper to fall at the same time.

4. Observe the position of the coin and paper as they fall.

NOTE: Keep the coin and the paper circle for the next experiment.

RESULTS The coin and paper fall together. They separate only after hitting the floor.

WHY? Falling objects speed up the same amount each second, thus the lighter paper and the heavier coin both fall at the same rate. Gravity at or near Earth causes the speed of falling objects to increase at a rate of

32 feet per second (9.8 meters per second) for every second of falling time. The heavier coin pushes through the air with more force than does the lightweight paper circle, but because their falling rate is the same and they are positioned on top of each other, they move downward as if they were one object. Any separation of the pair upon striking the floor is the result of their bouncing on the surface.

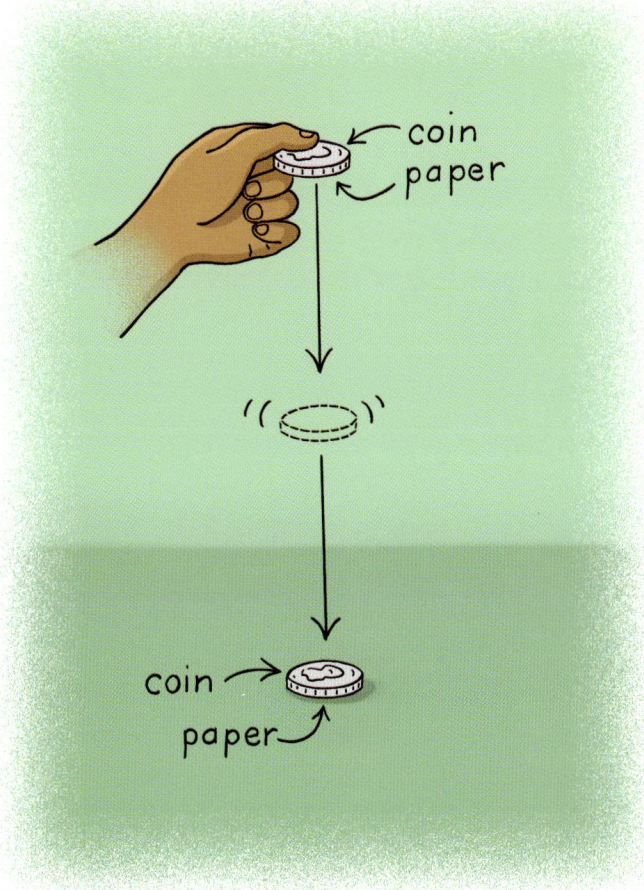

Reversed

Purpose To determine if changing the order of the falling coin and paper in the Paper Weight experiment would affect the results.

Materials coin and paper circle from previous experiment Paper Weight, see pp. 38–39
yardstick (meterstick)

Procedure

1. Hold the paper circle on top of the coin so that both are parallel with the floor. Be sure the paper does not extend past the edge of the coin.

2. Repeat steps 3 and 4 of the Paper Weight experiment, observing the position of the coin and paper as they fall.

Results The coin and paper fall together and separate only after hitting the floor.

Why? The heavier coin pushes through the air with more force than does the lightweight paper circle, and the two continue to fall at the same rate. When positioned on top of each other, they move downward as if they were one object regardless of which part of the pair is on top.

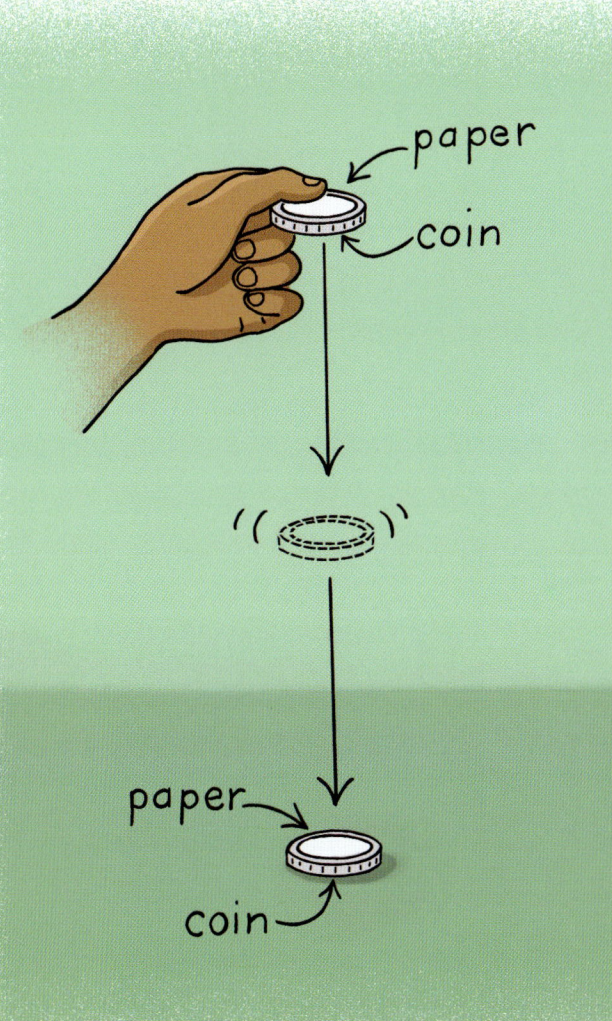

Lifter

Purpose To determine how air affects the falling rate of objects.

Materials pencil
coin
printer paper
scissors

Procedure

1. Use the pencil and the coin to draw a circle on the paper. Cut out the circle.

2. Holding the coin in one hand and the paper in the other about 3 feet (1 m) above the floor, release the coin and paper at the same time.

3. Observe the position of the coin and paper as they fall.

Results The coin falls straight down and the paper floats back and forth through the air. The coin hits the floor first.

Why? All things would fall on Earth at the same rate of 32 feet per second (9.8 mps) if there were no air pushing on them. But air molecules in Earth's atmosphere push against falling objects and slow their falling rate. Heavier objects, such as the coin, push through the air with more force than do lightweight objects, such as the paper. The air pushing on the lightweight paper lifts and slows its falling rate. Thus, heavier objects fall through air faster than do lightweight objects.

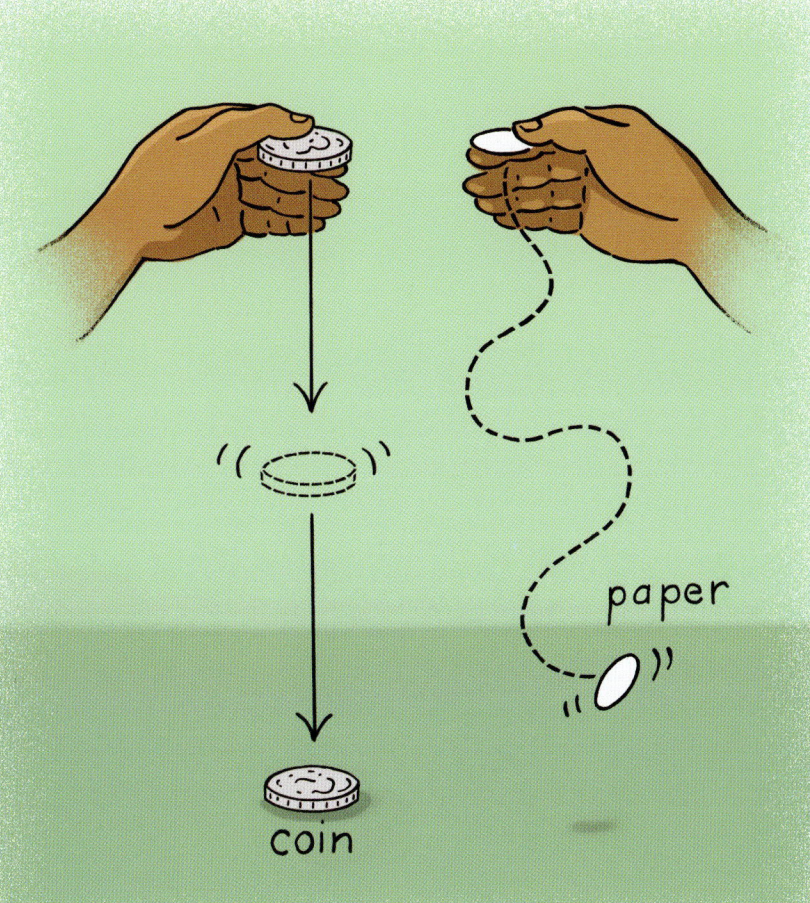

Lifter

Free-Fall

Purpose To determine the landing spot of free-falling objects.

Materials scissors
ruler
two 5-ounce (150 ml) paper cups
masking tape
yardstick (meterstick)
glass marble

Procedure

1. Cut one cup down to a height of about 1 inch (2.5 cm), then tape the cup to one end of the measuring stick.

2. Tape the other cup to the stick about 4 inches (10 cm) away from the first cup.

3. Tape the other end of the stick to the door frame. The stick must be loose enough to be raised up and down easily.

4. Place the marble in the cut cup.

5. Holding on to the stick about 8 inches (20 cm) from the free end (just behind the taller cup), raise the stick until the cup end is about 21 inches (53 cm) from the floor.

6. Allow the stick to fall to the floor. At the moment you release the stick, give it a gentle push downward.

7. Repeat steps 4 to 6 several times, each time changing the force of the push, until the following results are achieved.

RESULTS The marble moves out of the cut cup and falls into the taller cup. If the marble did not fall into the cup, adjust the downward force on the stick. Push a little harder if the marble falls short of the cup; decrease the force of the push if the marble moves past the taller cup.

WHY? Free-falling objects are pulled straight down toward Earth's center at a rate of 32 feet per second (9.8 mps) for every second of falling time. The push on the stick gives it a faster falling rate than the rate of free-falling objects. The faster-moving stick pulls the cup out from under the marble. The unsupported marble free-falls toward the floor. The path of the falling stick places the taller cup under the falling marble.

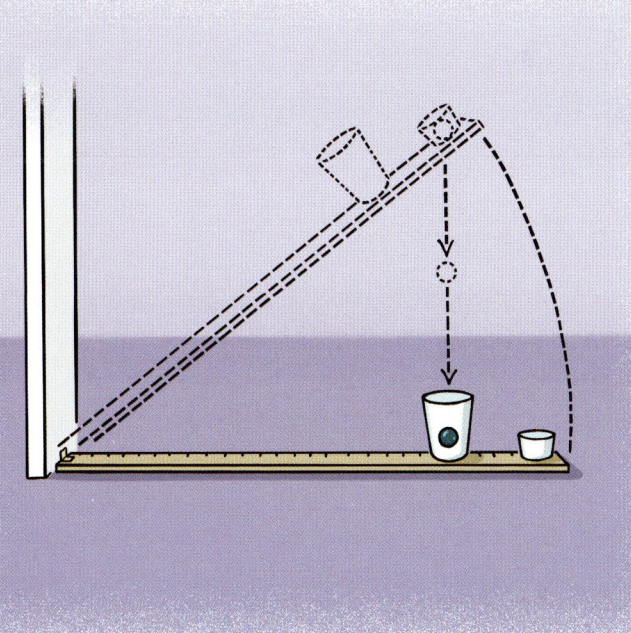

Free-Fall

Magic Box

Purpose To demonstrate center of gravity.

Materials rock
              small shoe box

Procedure

1. Place the rock in the inside corner of the shoe box, and close the lid.
2. Set the corner of the box that has the rock in it on the edge of a table. Be sure that no part of the rock is over the table's edge.

NOTE: If the box is difficult to balance, use a larger rock.

Results The box balances. Most of it is suspended over the edge of the table.

Why? An object's center of gravity is the point at which the object's weight seems to be concentrated so that the object would balance if supported at this point. The weight of the rock causes the center of gravity of the box and its contents to be on the tabletop. Therefore, the box balances on the table's edge.

Magic Box

Balancing Point

Purpose To demonstrate balance.

Materials two 10-inch (25 cm) pieces of string
lemon-size piece of clay
ruler

Procedure

1. Make a loop in each string by tying the ends together in a knot. Divide the clay into three equal parts. Squeeze one clay piece around the knotted end of each loop. Keep the third clay piece for a later step.

2. Place the loops on the ruler 2 inches (5 cm) in from each end.

3. Lay the ruler across your finger and move the ruler until it balances. Observe where your finger touches the ruler.

4. Add the third piece of clay to the clay on one of the loops.

5. Repeat step 4, observing where your finger touches the ruler this time.

Results At first, the ruler balances when your finger is at or near the middle of the ruler. When the extra clay was placed on one end, the ruler balanced when your finger was nearer the end with the larger clay piece.

Why? The place where the ruler balances on your finger is the center of gravity, which is the pivot point (the place around which an object turns). The downward force (weight) of each hanging clay piece pulls the ends of the ruler, causing them to turn in opposite directions, one clockwise and the other counterclockwise. Torque is the measure of the turning effect of each force and is determined by multiplying the force times its torque arm (the distance from the force to the pivot point). When the ruler balances, the torque on one side of the pivot point is equal to the torque on the opposite side.

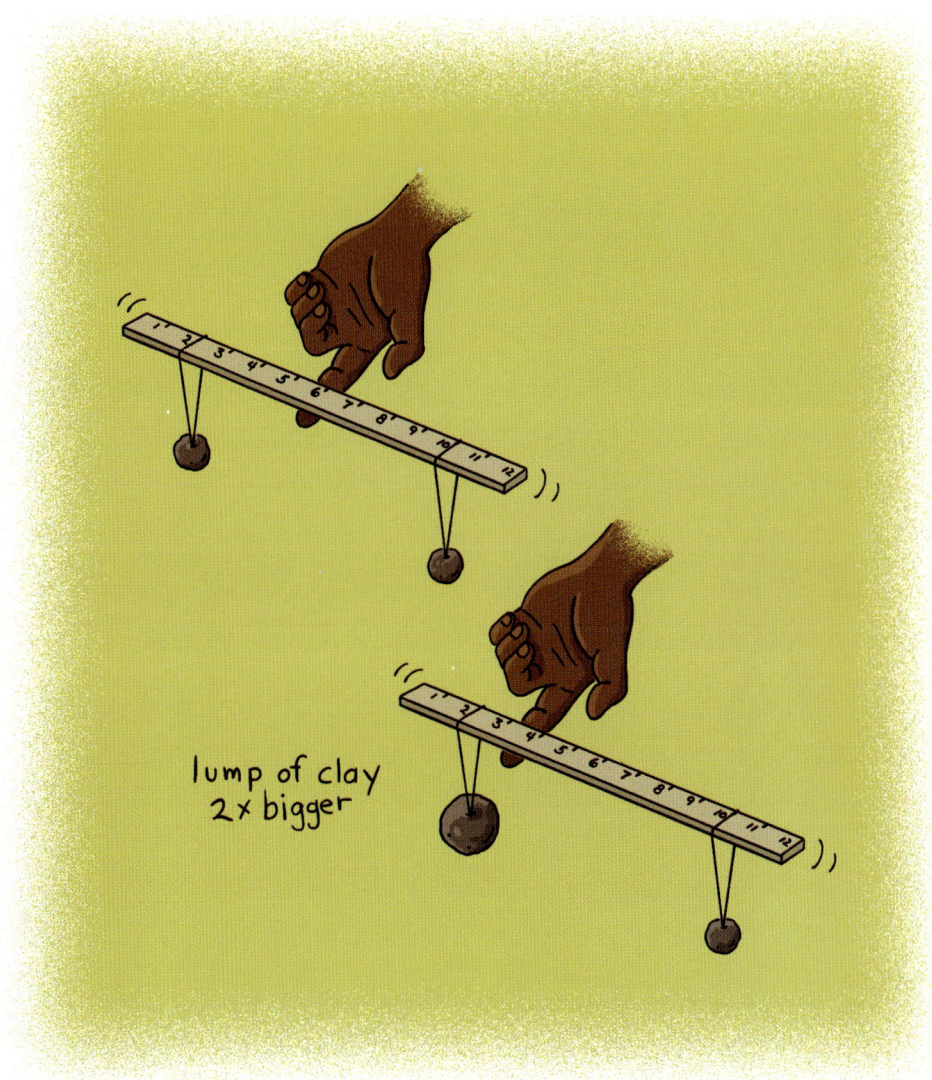

lump of clay 2× bigger

Unbalanced

Purpose To demonstrate the effect of equal but opposite forces.

Materials sharpened pencil

4-by-6-inch (10 x 15 cm) piece of corrugated cardboard
2 paper brads, or fasteners
rubber band
4 round pens

Procedure

1. Use the pencil to make two holes through the cardboard, one at either end of one short side.

2. Secure a paper brad in each hole, then wrap the rubber band around the paper brads as shown.

3. Place the pens parallel to each other on a flat surface, such as a table or floor. Then set the cardboard on the pens, separating the pens as much as possible beneath the cardboard.

4. Hold the cardboard with one hand. Stretch the rubber band toward the short side opposite the brads, forming a triangle as shown. Then release the cardboard and the rubber band. Observe what happens.

Results The rubber band moves forward and the cardboard moves backward.

Why? The motion of the cardboard can be explained by Newton's third law of motion, which states that for every action there is an equal and opposite reaction. When the stretched rubber band is released, it exerts an action force (a force applied to an object) on the cardboard. At the same time, the cardboard exerts a reaction force, which is a force equal to an action force, but in the opposite direction. When the rubber band is released, its action force is directed forward and the cardboard's reaction force pushes the cardboard in the opposite direction, or backward. These two forces are equal in magnitude, but opposite in direction.

BALLOON ROCKET

PURPOSE To demonstrate how a rocket moves.

MATERIALS 12-foot (3.6 m) or longer piece of string
4-inch (10 cm) piece of soda straw
9-inch (22.5 cm) round balloon
masking tape
ruler
helper

PROCEDURE

1. Thread the string through the straw.
2. Stretch the string across a room and attach the ends to the walls so that the string is taut and about waist high.
3. Inflate the balloon, squeeze it closed, and hold it under the straw.
4. Ask a helper to tape the balloon to the straw using two pieces of tape, each about 2 inches (5 cm) long.
5. Slide the straw and the inflated balloon to the end of the string that is nearer the mouth of the balloon.
6. Release the balloon and observe its motion.

RESULTS The balloon and straw move along the string as the balloon deflates.

WHY? The movement of the balloon, like that of a rocket, can be explained by Newton's third law of motion, which states that for every action force there is a reaction force. In the case of the balloon, the rubber pushes on the air inside, forcing it out the opening (an action force). The force of the air on the balloon (a reaction force) pushes the balloon in the opposite direction. Since the forces act on different materials (rubber and air), they don't cancel each other out. The same action-reaction forces move a rocket forward. The rocket pushes the gases downward (an action force) and the gases push the rocket upward (a reaction force).

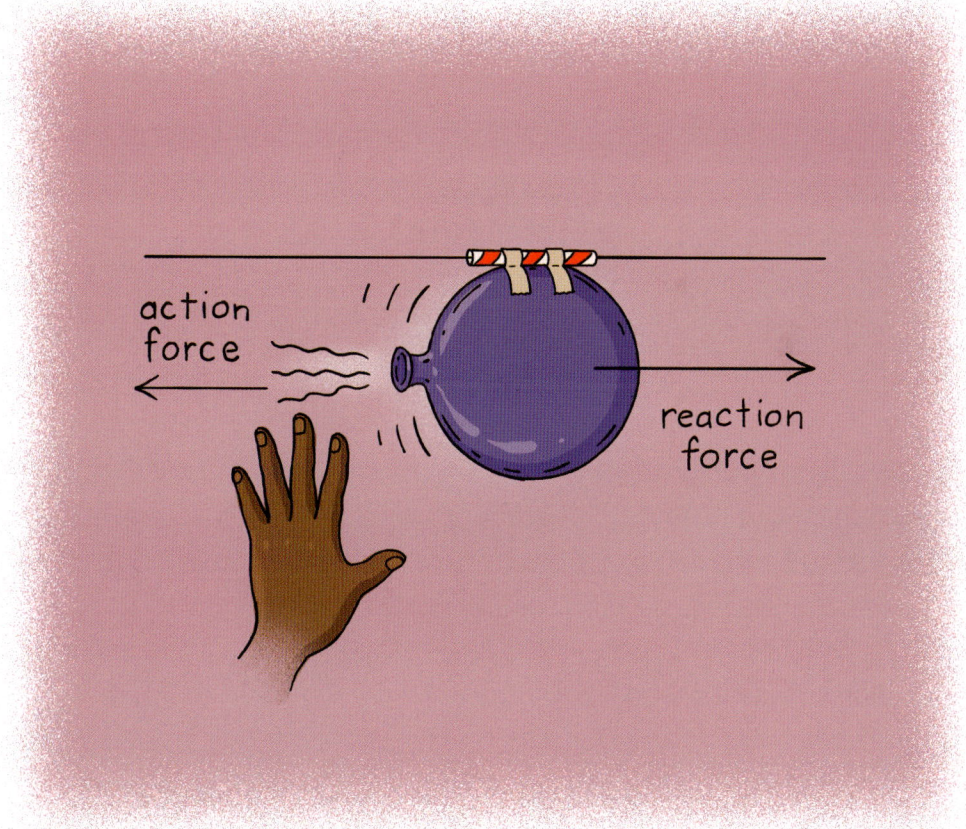

Balloon Rocket

Dancers

Purpose To demonstrate how the buoyancy of a material can be changed.

Materials 1 ½ cups (375 ml) white vinegar
1-pint (500 ml) jar
2 quarter-size pieces of eggshell from a boiled egg

Procedure

1. Pour the vinegar into the jar.

2. Break the eggshells into eight different-size pieces.

3. Drop the eggshells one at a time into the jar.

4. Observe what happens to the eggshells immediately and for the next few minutes.

Results Bubbles quickly form on the eggshells. Within two to three minutes, the shells start to rise and then sink in the liquid.

Why? Eggshells contain the chemical calcium carbonate. When calcium carbonate and vinegar mix, a chemical change occurs. One of the new chemicals produced is carbon dioxide gas. That's what is in the bubbles seen on the eggshells. These bubbles act like little life preservers and cause the shells to be buoyant (able to float in or on the surface of a fluid). As the shells rise, some of the bubbles get knocked away and the shell loses its buoyancy and sinks again.

55

Dancers

GLOSSARY

BUOYANCY The upward force exerted by a liquid such as water on any object in or on the liquid.

CENTER OF GRAVITY Point at which an object balances.

EFFORT FORCE The push or pull needed to move an object.

FIRST-CLASS LEVER A lever in which the fulcrum is between the two ends of the bar.

GRAVITY A force that pulls toward the center of a celestial body, such as Earth.

INCLINED PLANE A slanting or sloping surface used to raise an object to a higher level.

INERTIA A property of matter that causes objects to resist any change in motion.

LEVER A machine made of a rigid bar that turns around a pivot point called a fulcrum and is used to lift or move an object called the load.

MACHINE A device that makes work easier by changing the speed, direction, or amount of effort force applied.

MOLECULE The smallest particle of a substance; made of one or more atoms.

PITCH The property of sound that makes it high or low; also, the distance between the ridges winding around a screw.

SECOND-CLASS LEVER A lever in which the load is between the fulcrum and the effort force.

SIMPLE MACHINE A lever, inclined plane, wheel and axle, screw, wedge, or pulley.

SOUND A form of wave motion produced when objects vibrate.

THIRD-CLASS LEVER A lever in which the effort force is always greater than the load force.

VIBRATE To move quickly back and forth.

WEIGHT The downward pull that gravity has on an object.

For More Information

American Physics Society
  1 Physics Ellipse
  College Park, MD 20740
  (301) 209-3200
  Website: http://www.aps.org
  Facebook: @apsphysics
  Twitter: @APSphysics
  YouTube: @APS Physics
  Solve a mystery by doing an experiment through PhysicsQuest, read This Month in Physics History, or learn about careers in physics.

Canadian Association of Physicists
  555 King Edward Avenue
  3rd Floor
  Ottawa, ON K1N 7N5
  Canada
  (613) 562-5614
  Website: http://www.cap.ca
  Facebook: @CanadianAssociationOfPhysicists
  Twitter: @CAPhys
  Find out about careers in physics, enter the Art of Physics photography contest, or learn about student scholarships and prizes.

Intel
  2200 Mission College Boulevard
  Santa Clara, CA 95054-1549
  (408) 765-8080

Website: http://www.intel.com
Facebook: @Intel
Twitter: @intel
Instagram: @intel
YouTube: @Intel
Read student profiles of winning research projects from the Intel International Science and Engineering Fair, and find educational material about women in science, tips for your science fair project, and links to other competitions.

National Science Foundation (NSF)
4201 Wilson Boulevard
Arlington, VA 22230
(703) 292-5111
Website: http://www.nsf.gov
Facebook: @US.NSF
Twitter: @NSF
YouTube: @National Science Foundation
The NSF is dedicated to science, engineering, and education. Learn how to be a citizen scientist and read about scientific discoveries and innovations.

Society for Science and the Public
Student Science
1719 N Street NW
Washington, DC 20036
(800) 552-4412
Website: http://student.societyforscience.org
Facebook: @societyforscience
Twitter: @Society4Science

For More Information

Instagram: @society4science
YouTube: @societyforscience
The Society for Science and the Public presents science resources, such as science news for students, updates on the Intel Science Talent Search and the Intel International Science and Engineering Fair, and information about cool jobs and doing science.

USA Science & Engineering Festival
 Walter E. Washington Convention Center
 801 Mt. Vernon Place NW
 Washington, DC 20001
 (202) 459-0880
 Website: http://www.usasciencefestival.org
 Facebook: @usasciencefestival
 Twitter: @USAScienceFest
 YouTube: @USA Science & Engineering Festival
 The USA Science & Engineering Festival's mission is to advance STEM education and inspire the next generation of scientists and engineers. Nationwide school programs, contests, and events year-round culminate in a two-day Grand Finale Expo, free of charge.

Websites

Because of the changing nature of internet links, Rosen Publishing has developed an online list of websites related to the subject of this book. This site is updated regularly. Please use this link to access the list:

http://www.rosenlinks.com/JVCW/Phys

For Further Reading

Biskup, Agnieszka. *Super Cool Forces and Motion Activities with Max Axiom*. North Mankato, MN: Capstone Press, 2015.

Brown, Jordan. *Science Stunts: Fun Feats of Physics*. Watertown, MA: Charlesbridge Publishing, 2016.

Buczynski, Sandy. *Designing a Winning Science Fair Project*. Ann Arbor, MI: Cherry Lake Publishing, 2014.

Gardner, Robert. *The Physics of Sports Science Projects*. Berkeley Heights, NJ: Enslow Publishers, Inc., 2013.

Henneberg, Susan. *Creating Science Fair Projects with Cool New Digital Tools*. New York, NY: Rosen Central, 2014.

Kenney, Karen Latchana. *The Science of Music: Discovering Sound*. Minneapolis, MN: ABDO Publishing, 2016.

Latta, Sara L. *SMASH! Exploring the Mysteries of the Universe with the Large Hadron Collider*. Minneapolis, MN: Graphic Universe, 2017.

McCarthy, Cecilia Pinto. *The Science of Music*. Minneapolis, MN, Core Library, 2017.

Mercer, Bobby. *Junk Drawer Physics: 50 Awesome Experiments That Don't Cost a Thing*. Chicago, IL: Chicago Review Press, Inc., 2014.

Miller, Rachel. *The 101 Coolest Simple Science Experiments: Awesome Things to Do with Your Parents, Babysitters, and Other Adults*. Salem, MA: Page Street Publishing Co., 2016.

FOR FURTHER READING

O'Quinn, Amy M. *Marie Curie for Kids: Her Life and Scientific Discoveries, with 21 Activities and Experiments*. Chicago, IL: Chicago Review Press, 2017.

Rompella, Natalie. *Experiments in Light and Sound with Toys and Everyday Stuff*. North Mankato, MN: Capstone Press, 2016.

Sohn, Emily. *Experiments in Forces and Motion with Toys and Everyday Stuff*. North Mankato, MN: Capstone Press, 2016.

INDEX

A
action force, 50–51, 52–53
astronauts, 33, 34–35

B
balance, 46, 48–49, 50–51
buoyancy, 54

C
carbon dioxide, 54
center of gravity, 46, 48

D
data, 7

E
effort arm, 24
effort force, 18, 19, 20, 21, 22, 24, 27, 29, 30
Einstein, Albert, 4
energy, 4
engineers, 4

F
first-class lever, 22, 24, 27
flute, 8
forces, 4, 12, 14, 17, 18, 19, 20, 21, 22, 24, 26–27, 28–29, 30, 36, 39, 40, 42, 44, 45, 48–49, 50–51, 52–53
friction, 11
fulcrum, 22, 24, 26–27, 29

G
gravity, 4, 17, 19, 33, 34–35, 36, 38–39, 46, 48

H
hypothesis, 7

I
inclined plane, 20
inertia, 12–13, 14

L
lever, 22, 24, 26–27, 28–29
  first-class, 22, 24, 27
  second-class, 26–27
  third-class, 28–29
load, 22, 24, 26–27, 29, 30
load arm, 24

M
machine, 4, 20, 21, 22, 26, 30
mass, 17
matter, 4, 12

INDEX

molecules, 11, 42
motion, 12, 50, 52–53

N
Newton's third law of motion, 50, 52–53

P
physicists, 4
physics, 4, 6
pitch, 8
pivot point, 22, 48–49
pulley, 30

R
reaction force, 51, 52–53
rocket, 4, 52–53

S
safety, 5
scientific method, 6
second-class lever, 26–27
simple machine, 4, 20, 26, 30
sound, 8, 10, 11
sound waves, 11

T
third-class lever, 28–29
torque, 48–49

V
vibration, 8, 10, 11

W
weight, 17, 19, 33, 38, 39, 40, 42, 46, 48
weightlessness, 32, 33